Busy Ant Maths

Problem Solving and Reasoning Pupil Book 5

Peter Clarke

William Collins' dream of knowledge for all began with the publication of his first book in 1819. A self-educated mill worker, he not only enriched millions of lives, but also founded a flourishing publishing house. Today, staying true to this spirit, Collins books are packed with inspiration, innovation and practical expertise. They place you at the centre of a world of possibility and give you exactly what you need to explore it.

Collins. Freedom to teach.

Published by Collins
An imprint of HarperCollins*Publishers*
The News Building
1 London Bridge Street
London
SE1 9GF

Browse the complete Collins catalogue at
www.collins.co.uk

© HarperCollins*Publishers* Limited 2018

10 9 8 7 6 5 4 3 2 1

ISBN 978-0-00-826050-7

The author wishes to thank Brian Molyneaux for his valuable contribution to this publication.

British Library Cataloguing in Publication Data
A Catalogue record for this publication is available from the British Library

Author: Peter Clarke
Publishing manager: Fiona McGlade
Editor: Amy Wright
Copyeditor: Catherine Dakin
Proofreader: Tanya Solomons
Answer checker: Steven Matchett
Cover designer: Amparo Barrera
Internal designer: 2hoots Publishing Services
Typesetter: Ken Vail Graphic Design
Illustrator: Eva Sassin
Production controller: Sarah Burke
Printed and bound by Martins the Printers

Contents

Contents

How to use this book

Aims

This book aims to provide teachers with a resource that enables pupils to:

- develop mathematical problem solving and thinking skills
- reason and communicate mathematically
- use and apply mathematics to solve problems.

The three different types of mathematical problem solving challenge

This book consists of three different types of mathematical problem solving challenge:

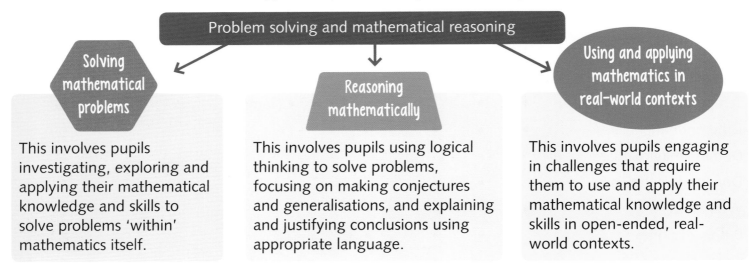

This involves pupils investigating, exploring and applying their mathematical knowledge and skills to solve problems 'within' mathematics itself.

This involves pupils using logical thinking to solve problems, focusing on making conjectures and generalisations, and explaining and justifying conclusions using appropriate language.

This involves pupils engaging in challenges that require them to use and apply their mathematical knowledge and skills in open-ended, real-world contexts.

This book is intended as a 'dip-in' resource, where teachers choose which of the three different types of challenge they wish pupils to undertake. A challenge may form the basis of part of or an entire mathematics lesson. The challenges can also be used in a similar way to the weekly bank of 'Learning activities' found in the *Busy Ant Maths* Teacher's Guide. It is recommended that pupils have equal experience of all three types of challenge during the course of a term.

The 'Solving mathematical problems' and 'Reasoning mathematically' challenges are organised under the different topics (domains) of the 2014 National Curriculum for Mathematics. This is to make it easier for teachers to choose a challenge that corresponds to the topic they are currently teaching, thereby providing an opportunity for pupils to practise their pure mathematical knowledge and skills in a problem solving context. These challenges are designed to be completed during the course of a lesson.

The 'Using and applying mathematics in real-world contexts' challenges have not been organised by topic. The very nature of this type of challenge means that pupils are drawing on their mathematical knowledge and skills from several topics in order to investigate challenges arising from the real world. In many cases these challenges will require pupils to work on them for an extended period, such as over the course of several lessons, a week or during a particular unit of work. An indication of which topics each of these challenges covers can be found on page 5.

Briefing

As with other similar teaching and learning resources, pupils will engage more fully with each challenge if the teacher introduces and discusses the challenge with the pupils. This includes reading through the challenge with the pupils, checking prerequisites for learning, ensuring understanding and clarifying any misconceptions.

Working collaboratively

The challenges can be undertaken by individuals, pairs or groups of pupils, however they will be enhanced greatly if pupils are able to work together in pairs or groups. By working collaboratively, pupils are more likely to develop their problem solving, communicating and reasoning skills.

You will need

All of the challenges require pupils to use pencil and paper. Giving pupils a large sheet of paper, such as A3 or A2, allows them to feel free to work out the results and record their thinking in ways that are appropriate to them. It also enables pupils to work together better in pairs or as a group, and provides them with an excellent prompt to use when sharing and discussing their work with others.

An important problem solving skill is to be able to identify not only the mathematics, but also what resources to use. For this reason, many of the challenges do not name the specific resources that are needed.

Characters

The characters on the right are the teacher and the four children who appear in some of the challenges in this book.

Mr Menel

Isabel

Georgia

Joshua

Xavier

Think about …

All challenges include prompting questions that provide both a springboard and a means of assisting pupils in accessing and working through the challenge.

What if?

The challenges also include an extension or variation that allows pupils to think more deeply about the challenge and to further develop their thinking skills.

When you've finished, …

At the bottom of each challenge, pupils are instructed to turn to page 80 and to find a partner or another pair or group. This page offers a structure and set of questions intended to provide pupils with an opportunity to share their results and discuss their methods, strategies and mathematical reasoning.

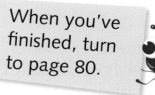
When you've finished, turn to page 80.

Solutions

Where appropriate, the solutions to the challenges in this book can be found at *Busy Ant Maths* on Collins Connect and on our website: collins.co.uk/busyantmaths.

Swap to order

Solving mathematical problems

Challenge

Shuffle a set of 0–9 digit cards.

Place the 10 cards, face up, in a row.

For example:

8 3 6 0 4 5 1 7 2 9

Complete the statement below by swapping two cards.

☐☐☐☐☐ < ☐☐☐☐☐

For example:

5 3 6 0 4 < 8 1 7 2 9

Now complete the statement below by swapping two cards. The two numbers must be different from the numbers you swapped in the previous statement.

☐☐☐☐☐ > ☐☐☐☐☐

For example:

9 3 6 0 4 > 8 1 7 2 5

Repeat the above process twice more until you have written six different statements: three using the 'less than' symbol, and three using the 'greater than' symbol.

Record all the statements you make.

Then order your 12 numbers, from smallest to largest.

You will need:
- set of 0–9 digit cards
- any two additional 0–9 digit cards, for example 4 and 7

Think about ...

Remember, you can't use the same number twice. All 12 numbers must be different.

Remember, you're alternating between statements, using the 'less than' and 'greater than' symbols.

What if?

What if you use 12 digit cards and complete the statements below?

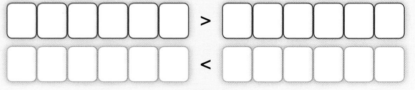

☐☐☐☐☐☐ > ☐☐☐☐☐☐

☐☐☐☐☐☐ < ☐☐☐☐☐☐

Write six different statements: three using the 'greater than' symbol, and three using the 'less than' symbol.

Record your statements, then order the 12 numbers, starting with the smallest.

When you've finished, turn to page 80.

What are the number sequences?

Challenge

235 500   75   2750   215 150   24 150   265 150   4950
235 200   7950   −75   245 150   235 300
235 000   225 150   3950   29 150   235 600
25 150   2350   235 400   26 150
−50   2650   0   9950   2450
6950   235 400   −25
2550   27 150   2950   2850
25   23 150   5950   235 100
205 150   8950
255 150   235 150   50
28 150

These 42 numbers are from six different number sequences. Each sequence has seven numbers.

Identify the six number sequences, writing the numbers in order, from smallest to largest.

Then for each sequence, look at the seven numbers and write the number that comes before the first number, and the number that comes after the last number.

Think about …

Which digits are the same among certain sets of numbers? Which digits are different? What does this tell you?

What patterns can you spot in the different sets of numbers?

What if?

What is the rule for each sequence?

Using those same six rules, write another six similar number sequences. Make sure each sequence has seven numbers.

Rewrite your list of 42 numbers so that they are all mixed up and give them to a friend. Can they identify your six number sequences, writing the numbers in order from smallest to largest?

When you've finished, turn to page 80.

Challenge

Roll a 1–6 dice five times. After each roll of the dice, write down the number.

Using your five numbers, make as many 5-digit numbers between 40 000 and 60 000 as you can.

Round each number to the nearest 10, 100, 1000 and 10 000.

Now sort your numbers to show which numbers round to the same 10, 100, 1000 or 10 000.

> **You will need:**
> • 1–6 dice

Think about ...

The number of 5- and 6-digit numbers you can make will depend on the numbers you roll. Don't make too many numbers – around 30 is enough.

Think carefully about the best way to record how you sorted your numbers.

What if?

What if you roll the dice six times?

How many different 6-digit numbers can you make between 300 000 and 500 000?

Round each number to the nearest 10, 100, 1000, 10 000 and 100 000.

Then sort your numbers to show which numbers round to the same 10, 100, 1000, 10 000 or 100 000.

When you've finished, turn to page 80.

Challenge

Step ①: Choose a pair of 3-digit numbers.

Step ②: Mentally subtract the smaller number from 999.

Step ③: Add this answer to the larger 3-digit number you chose.

Step ④: Look at your 4-digit answer. Separate the thousands digit from the hundreds, tens and ones digits to get a 1-digit and a 3-digit number.

Step ⑤: Add together the 1-digit number and the 3-digit number.

Now go back to the pair of 3-digit numbers you originally chose. Find the difference between these two numbers.

What do you notice?

Does this method always work?

Think about ...

Do as many of the calculations as you can mentally.

For the 'What if?', think carefully about **Steps ②** and **④**.

What if?

What if you choose a pair of 4-digit numbers?

What modifications do you need to make to the five steps above?

When you've finished, turn to page 80.

Challenge

Use the digits shown on the right to make 12 different 5-digit numbers between 59 000 and 63 000.

Write the numbers in order, from smallest to largest.

Find the pair of numbers with:

- the smallest total
- the greatest total
- the smallest difference
- the greatest difference.

Write about how you knew which pairs of numbers make the smallest and greatest totals and differences.

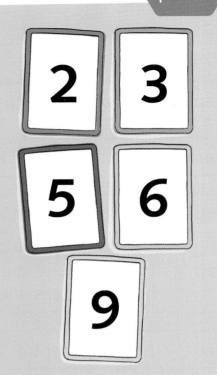

Think about ...

Think carefully about which numbers to choose that will give you the smallest and greatest totals and differences.

For the 'What if?', there is more than one pair of numbers that have the smallest difference. How many different pairs of numbers can you find?

What if?

Altogether it's possible to make 120 different 5-digit numbers using the digits above.

Without making all 120 numbers, work out which pair of numbers has:

- the smallest total
- the greatest total
- the smallest difference
- the greatest difference.

Use what you discovered in the 'Challenge' to help you choose the correct pairs of numbers.

When you've finished, turn to page 80.

Challenge

Step 1.
Write down a 3-digit number.

Step 2.
Reverse the digits in the number.

Step 3.
Add the two numbers together.

STOP

YES

Step 4 .
Is the answer a palindromic number?

Repeat steps 2, 3 and 4.

NO

> The digits in a palindromic number read the same backwards as forwards.

Repeat the steps above, starting with different 3-digit numbers.

Which numbers become palindromic after one addition?

Which numbers become palindromic after two additions? Three additions? Four additions?

Think about ...

> Look carefully before you carry out each addition and decide whether to use a mental strategy or a written method.

> Don't try 196. Many mathematicians have used computers and, even after several thousand additions, they still can't come up with a palindromic number.

What if?

What if you use 4-digit numbers?

When you've finished, turn to page 80.

Challenge

The factors of
16
are
**1, 2, 4, 8
and 16**

The factors of
15
are
**1, 3, 5
and 15**

Investigate which 2-digit numbers have the most factors.

What do you notice about these numbers?

Think about ...

Think carefully about which 2-digit numbers are likely to have the most factors – start with these numbers.

Think about the relationship between different numbers and how these numbers might have common factors.

What if?

Which 2-digit numbers have the fewest factors?

What do you know about these numbers?

When you've finished, turn to page 80.

Challenge

I can make the number 45 by adding two square numbers.

$$45 = 6^2 + 3^2$$

Investigate which numbers from 1 to 50 you can make by adding and/or subtracting square numbers.

Think about ...

Think about whether you need to add or subtract two or more square numbers.

Try to perform just one or two operations for each number.

Use all the square numbers up to 12^2.

Look at the calculations you have used to make different numbers and see how these calculations might help you to make other numbers.

What if?

What numbers can you make by adding and/or subtracting combinations of squared (^{2}) and cubed (^{3}) numbers?

When you've finished, turn to page 80.

Challenge

Shuffle a set of 1–9 digit cards.

Deal the top four cards and place them face up on the table.

Arrange the four digits in the calculation below.

☐ • ☐ ☐ ✗ ☐ =

How many different calculations can you make by rearranging your four digit cards?

You will need:
• set of 1–9 digit cards

Think about ...

Think about what changes and what stays the same when you rearrange your four digit cards to make all the different calculations possible.

Look carefully at each calculation, and think about the best strategy to use to work out the answer.

What if?

What if you deal five cards and you arrange them in the calculation below?

☐ • ☐ ☐ ☐ ✗ ☐ =

With this arrangement, you could make 120 different calculations!

Don't make 120 calculations – just make the same number as you did in the 'Challenge' above.

When you've finished, turn to page 80.

Mixed numbers

Challenge

Write a mixed number that lies between each pair of labelled numbers and fractions on these number lines.

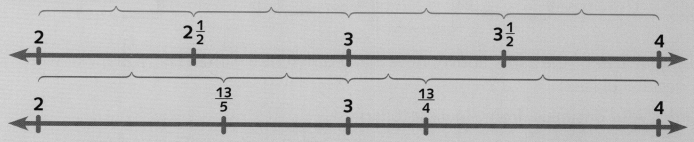

Can you think of a second mixed number that lies between each pair of labelled numbers and fractions?

Think about ...

All 16 of your mixed numbers must be different and you can't write the same fraction twice. For example: $2\frac{1}{4}$ and $3\frac{1}{4}$ ✗ (You can't write $\frac{1}{4}$ more than once.)

Think carefully about what fractions to use in your mixed numbers. Using a **fraction family**, with the same denominators or denominators that are all multiples of the same number, will help you.

What if?

Use the 16 mixed numbers you have written on the two number lines to complete these statements. You can only use each mixed number once.

☐ < ☐   ☐ < ☐ > ☐   ☐ > ☐ > ☐

☐ > ☐   ☐ < ☐ < ☐   ☐ > ☐ < ☐

Look at your completed statements above. Rewrite each statement, converting mixed numbers to improper fractions.

When you've finished, turn to page 80.

Challenge

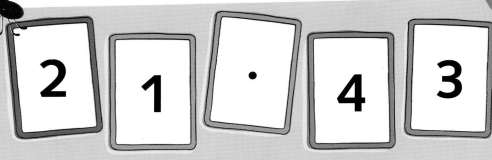

Using all five of the cards above, how many different decimals can you make with **1** decimal place?

Order your decimals, from smallest to largest.

How many different decimals can you make with **2** decimal places? Order your decimals, smallest first.

How many decimals can you make with **3** decimal places? Order them.

Think about ...

Look for patterns between the decimals you can make with 1, 2 and 3 decimal places.

For the 'What if?', make decimals with 1, 2 and 3 decimal places.

What if?

Using some or all of these cards, which decimals can you make that are less than $\frac{1}{2}$?

Write your decimals in order, smallest first.

Which decimals can you make that are greater than $\frac{3}{4}$ and less than 1 using some or all of these cards?

Order your decimals, smallest first.

When you've finished, turn to page 80.

Challenge

Investigate what percentage of the numbers on a 1 to 100 number square are:

- odd numbers
- 1-digit numbers
- 2-digit numbers
- 3-digit numbers
- multiples of 5
- multiples of 2 **and** 5
- multiples of 3
- numbers that contain the digit 7
- square numbers
- factors of 36
- prime numbers.

1	2	3	4	5	6	7	8	9	10
11	12	13	14	15	16	17	18	19	20
21	22	23	24	25	26	27	28	29	30
31	32	33	34	35	36	37	38	39	40
41	42	43	44	45	46	47	48	49	50
51	52	53	54	55	56	57	58	59	60
61	62	63	64	65	66	67	68	69	70
71	72	73	74	75	76	77	78	79	80
81	82	83	84	85	86	87	88	89	90
91	92	93	94	95	96	97	98	99	100

Think about ...

Remember that 'per cent' relates to the number of parts per hundred, and there are 100 numbers, or parts, on a 1 to 100 number square.

For the 'What if?', think about place value, comparing, ordering and rounding numbers, and factors and multiples. What other topics can you think of?

What if?

Investigate different types of numbers on a 1 to 100 number square that are 25% of the square.

What about 60%? What about 12%?

Find other types of numbers that are different percentages of a 1 to 100 number square.

When you've finished, turn to page 80.

19

Challenge

1	2	4	5	10	20

2	4	5	10	25	50

Choose a number from the **red cards** as the **numerator**, and a number from the **blue cards** as the **denominator** to make a proper fraction.

How many different proper fractions can you make?

Which of your fractions can you convert to a decimal?

Which can you convert to a percentage?

Think about ...

Think carefully about equivalent fractions, decimals and percentages.

Think about how to convert a fraction to a hundredth and how this will help you write the equivalent decimal and percentage.

What if?

Look at all the fractions you have made using the cards above.

How can you sort your fractions into different groups?

Can you sort your fractions in more than one way?

When you've finished, turn to page 80.

Challenge

Measure the length of your step to the nearest centimetre.

How many steps would it take you to travel 1 km? 10 km? 12 km? 25 km? 100 km?

If you could take 1 million steps, approximately how many kilometres would you travel?

Show all of your working out.

You will need:
- measuring equipment
- map or distance chart giving distances between various cities or landmarks

Think about ...

You don't need to come up with an exact number of steps or kilometres, so you might want to round the length of your step up or down slightly to give you a 'friendly' measurement to work with.

Once you have worked out how many steps it would take you to travel 1 km, use this to help find the number of steps you would take for the other distances.

What if?

Look at a map or distance chart (or use the one provided here) to find the distance between cities or landmarks. Choose two locations and find out the distance between them. Then work out approximately how many steps you would need to take to travel from one location to the other.

Birmingham	Bristol	Cardiff	Edinburgh	Glasgow	Nottingham	Liverpool	London
160							
190	72						
470	601	635					
466	597	631	75				
82	225	259	452	484			
160	292	326	358	354	179		
204	190	244	667	663	210	357	

Distances are in km

When you've finished, turn to page 80.

Expanding quadrilaterals

Solving mathematical problems

Challenge

Draw a rectangle. Label it **A** and work out its area and perimeter.

Draw shape **A** again, but with each side twice as long.

Label it **B** and work out its area and perimeter.

Now draw shape **A** again, but this time, with each side three times as long.

Label it **C** and work out its area and perimeter.

What do you notice about the areas and perimeters of shapes **A**, **B** and **C**?

You will need:
- 1 cm squared paper
- ruler

Think about ...

Draw the length of each side of your quadrilaterals to a whole number of centimetres.

Think carefully about the length of the sides of shape **A**. Remember, you need to double these dimensions for shape **B** and triple them for shape **C**, so you shouldn't start with too large a quadrilateral for shape **A**.

What if?

Draw a different quadrilateral and repeat the steps above, labelling each shape **A**, **B** and **C**.

What do you notice about the areas and perimeters of shapes **A**, **B** and **C**?

Try with other quadrilaterals.

When you've finished, turn to page 80.

Challenge

Each sack contains wheat, sugar or rice.

There is twice as much wheat as sugar, and only one sack contains rice.

Work out what each sack contains.

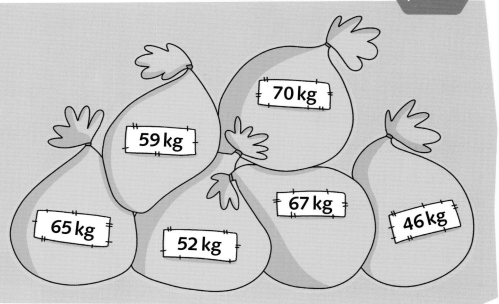

70 kg

59 kg

67 kg

65 kg

52 kg

46 kg

Think about ...

Begin by making estimates, and then use trial and improvement.

Show all your working out.

What if?

These sacks contain wheat, sugar, rice or corn.

There is three times as much wheat as sugar, and twice as much sugar as rice.

There is the same mass of rice as corn, but there is one more sack of rice than corn.

Work out what each sack contains.

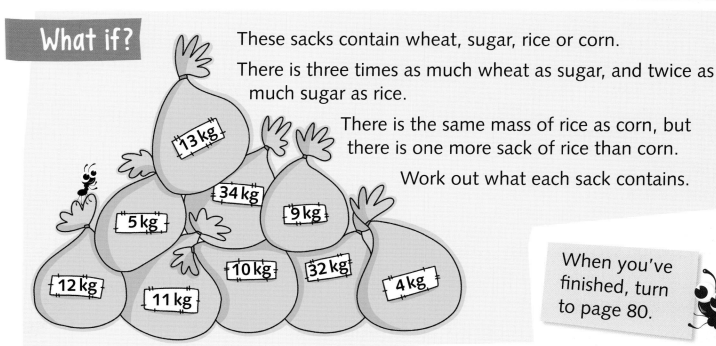

13 kg

34 kg

5 kg

9 kg

12 kg

10 kg

32 kg

11 kg

4 kg

When you've finished, turn to page 80.

When will it be?

Challenge

Write today's date.

Write the time to the nearest minute.

What time and date will it be:

- 100 minutes from now?
- 1000 minutes from now?
- 100 hours from now?
- 1000 hours from now?
- 100 days from now?
- 1000 days from now?

Show all your working out.

12th March 2018
10:15 a.m.

Think about ...

Think carefully about exactly when to start this challenge and what time to write down. Why might 10:15, rather than 10:13 or 10:14 be a better time to start?

For the 'What if?', you might find it too difficult to carry out the calculations. If so, just write down all the calculations you need to do but don't work out the answers. Draw shapes or letters to stand for the unknown numbers in your calculations.

What if?

How long is:

- 1 million seconds in days?
- 1 million minutes in weeks?
- 1 million hours in years?

Give your answers rounded to the nearest day, week and year.

When you've finished, turn to page 80.

Challenge

Write a number in the square to represent a percentage and then any of the coins in the circles to complete the statement.

☐ % of £ ◯ = ◯ p

You can only use each coin once for each statement.

Joshua has written a statement:

Investigate how many different statements you can make.

50% of £1 = 50p

Think about ...

Think carefully about which percentages are most suitable.

Show all your working out.

What if?

Would you be willing to take just one penny as pocket money this week, as long as each week after that the amount of money would double?

How much pocket money would you get in the 8th week?

What about the 12th week? 15th week? 20th week? 24th week?

Week 1: 1p
Week 2: 2p
Week 3: 4p
Week 4: 8p
Week 5:

When you've finished, turn to page 80.

Clock angles

Challenge

You will need:
• protractor

If an analogue clock reads 12:10, what is the approximate size of the angle formed by its hands?

What if it reads 2:25?

What about 4:45?

Choose other times to 5 minute intervals and work out the approximate size of each angle. Then name the angle.

Think about ...

Think carefully about whether you need to measure the size of the angles using a protractor or whether you can work it out.

You are thinking about the smaller angle for each time, for example:

not this angle

this angle

What if?

Choose times to 1 minute intervals, like 6:47, and work out the approximate size of the angles. Name each angle.

When you've finished, turn to page 80.

26

Angle polygons

Challenge

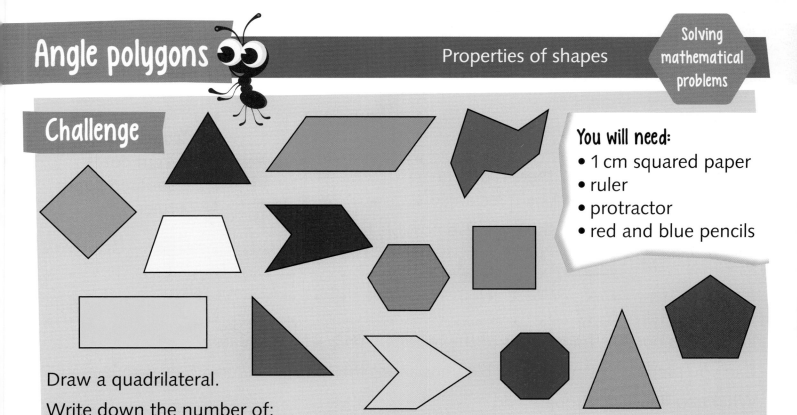

You will need:
- 1 cm squared paper
- ruler
- protractor
- red and blue pencils

Draw a quadrilateral.

Write down the number of:

- right angles
- acute angles
- obtuse angles
- reflex angles.

Now estimate the size of each interior angle. Write your estimates in red.

Then measure, or work out, the size of each interior angle. Write the actual size of each interior angle in blue.

What can you say about the sum of all the interior angles in different quadrilaterals?

Write about what you notice.

Think about ...

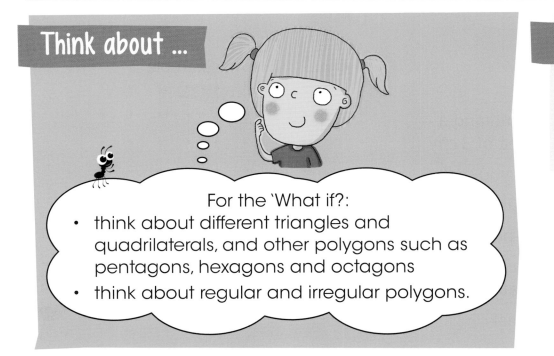

For the 'What if?':
- think about different triangles and quadrilaterals, and other polygons such as pentagons, hexagons and octagons
- think about regular and irregular polygons.

What if?

Investigate different triangles and other polygons.

When you've finished, turn to page 80.

Challenge

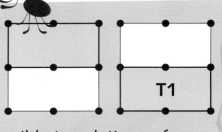

On a 2 by 2 square grid, there is one possible translation of a 1 by 2 rectangle.

You will need:
- 1 cm dot squared paper
- ruler

On a 3 by 3 square grid, there are five possible translations of a 1 by 2 rectangle.

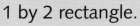 

Investigate possible translations of the rectangle on a 4 by 4 square grid.

Can you predict how many possible translations there would be on a 5 by 5 square grid?

What about a 6 by 6 square grid?

Explain how you arrived at your predictions.

Think about ...

What relationship do you notice between the increase in the size of the grid and the number of possible translations?

Work systematically in order to spot all the possible translations.

What if?

Start with a 3 by 3 square grid and a rectilinear shape with an area of 3 cm².

How many translations can you make?

Increase the size of the square grid.

How many translations are possible?

Repeat several times, increasing the size of the square grid each time and investigating possible translations.

What predictions can you make?

Example 1   Example 2

When you've finished, turn to page 80.

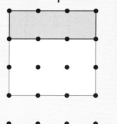

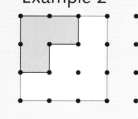

28

Reflect and translate

Challenge

Simple shapes can be made into a design by reflections or translations.

Reflections

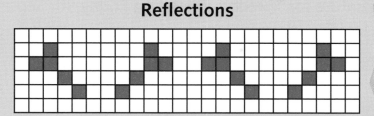

Translations

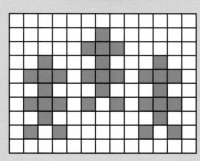

Using squared paper, make a simple shape by colouring several squares in the same colour.

Then reflect and/or translate your shape several times to create a design.

Write about how you created your design.

You will need:
- 0·5 cm or 1 cm squared paper
- coloured pencils

Think about ...

Colour between 5 and 10 squares to make your original simple shape.

If your design includes reflections, translations **and** rotations, make sure you can explain how you 'transformed' your original shape each time.

What if?

Designs can also be made using rotations.

Create a one-colour design using reflections and/or translations and/or rotations.

Rotations

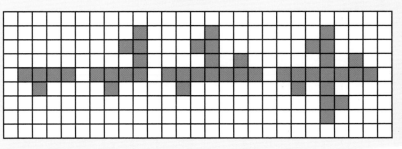

Create a design using reflections and/or translations and/or rotations and two or more colours.

When you've finished, turn to page 80.

A winter's day in Thredbo Village

Challenge

Thredbo Village is a ski resort in New South Wales, Australia, where the ski season runs from June to September.

You will need:
- squared or graph paper
- ruler

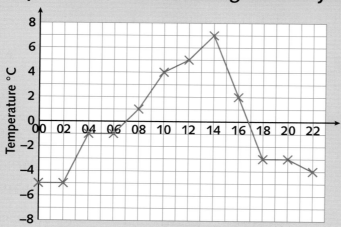

Temperature for Thredbo Village on Friday 8th September

What statements can you make about the temperatures in Thredbo Village on Friday 8th September?

Think about ...

Make statements describing and comparing hour-by-hour temperatures, and also the temperatures at different parts of the day: morning, afternoon, evening and night.

In your statements, include how the temperature rises and falls during the course of a day, and also by how many degrees.

What if?

Look at the table below showing the temperatures in Thredbo Village a month earlier on Tuesday 8th August.

Time	00	01	02	03	04	05	06	07	08	09	10	11	12	13	14	15	16	17	18	19	20	21	22	23
Temp (°C)	−6	−7	−7	−5	−3	−2	−1	−1	0	1	2	3	4	6	5	4	1	0	−2	−3	−3	−4	−5	−5

Draw a line graph similar to the one above for the temperatures in Thredbo Village on Tuesday 8th August.

What statements can you make comparing temperatures in Thredbo Village on these two days?

When you've finished, turn to page 80.

Timetables

Challenge

Birmingham International	07:24	09:27	11:23	13:23	14:23	15:24	16:24	17:26	18:41	19:22
Coventry	07:56	09:58	11:55	13:55	14:55	15:56	16:56	17:56	19:16	19:56
Leamington		10:09		14:06		16:07		18:06		20:06
Oxford	08:41	10:42	12:39	14:39	15:39	16:40	17:40	18:41	20:01	20:41
Reading	08:53	10:55	12:53	14:53	15:53	16:53	17:53	18:54	20:14	20:54
Basingstoke	08:58		12:58		15:58		17:58		20:19	
Southampton Airport	09:07	11:08	13:06	15:06	16:06	17:06	18:06	19:09	20:28	21:10
Southampton Central	09:38	11:39	13:37	15:37	16:37	17:37	18:34	19:36	20:56	21:41
Bournemouth	10:19	12:27	14:17	16:08	17:10	18:12	19:15	20:19	21:31	22:29

This is part of the train timetable for Birmingham International to Bournemouth for Monday to Friday.

When is the fastest train?

When is the slowest train?

To the nearest 5 minutes, approximately how long does the average trip take?

Which two consecutive stations are the furthest apart? Why do you think this?

Look at the timetable above and write five different statements.

Think about ...

Write statements about the times different trains take between Birmingham International and Bournemouth, as well as comparing the different train stops between Birmingham International and Bournemouth.

Some of your observations should be based on time differences.

What if?

Birmingham International	07:29	09:29	11:29	13:29	15:29	17:29	19:29
Coventry	08:01	10:00	12:00	14:00	16:00	18:00	19:59
Leamington	08:13	10:13	12:13	14:13	16:13	18:13	20:11
Oxford	08:46	10:48	12:48	14:48	16:48	18:48	20:53
Reading	08:59	11:01	13:01	15:01	17:01	19:05	21:13
Basingstoke	09:08	11:08	13:08	15:08	17:08	19:12	21:20
Southampton Airport	09:14	11:15	13:15	15:15	17:15	19:19	21:30
Southampton Central	09:47	11:47	13:47	15:47	17:47	19:51	21:57
Bournemouth	10:24	12:21	14:19	16:28	18:21	20:23	22:45

This is part of the train timetable for Birmingham International to Bournemouth for Saturday, Sunday and Public Holidays.

Look at the timetable above and write five different statements.

Make statements comparing the two timetables.

When you've finished, turn to page 80.

Challenge

When I count back from 12 in steps of 3, I will say "–3".

When I count back from 12 in steps of 5, I will say "–5".

When I count back from 12 in steps of 4, I will say "–4".

Who's correct?

Explain your reasoning.

Think about ...

Explain your reasoning both in words and using a diagram such as a number line.

Think carefully about what happens when you count through zero.

What if?

If each child counts from 14, what is the 6th number they will each say?

I'm counting back in steps of 5.

I'm counting back in steps of 4.

I'm counting back in steps of 6.

When you've finished, turn to page 80.

Challenge

Isabel counts forwards and backwards in steps of 10 from 3642.

Which of these numbers will Isabel say as she counts?

14 742	364	642	–2
7640	12	–8	64
8512	23 758	36 420	902

Explain how you know which numbers Isabel will count.

Think about ...

Think about which digits change and which digits remain the same when you count on or back in steps of 10, 100 or 1000.

Think carefully about which negative numbers Isabel will say.

What if?

What if Isabel counts forwards and backwards in steps of 100 from 64 781?

Which of these numbers will Isabel say as she counts? Explain how you know.

–19	64 791	478	193 881
6478	381	78	–81
4581	647 810	50 481	69 981

What if Isabel counts forwards and backwards in steps of 1000 from 371 497.

Which of these numbers will Isabel say as she counts? Explain how you know.

513 497	371 097	497	353 497
381 597	385 497	49	3714
– 497	2490	11 497	– 503

When you've finished, turn to page 80.

Challenge

Complete these calculations.

A   XI + XXIII

B   LVI – XXV

C   XIII × VIII

D   LXXII ÷ IX

E $\frac{I}{V}$ of XLV

Show your working out and explain how you found the answers.

Think about ...

Show how you translated the Roman numerals into Hindu-Arabic numerals.

Write your answers in both Roman numerals and Hindu-Arabic numerals.

What if?

Complete these calculations.

F   DCXLVII + CCCLXXXV

G   MMMMDCLII – MCMXLVI

H   LXIII × LVIII

I   MCMXLIV ÷ VI

J $\frac{V}{VIII}$ × XCVI

Show your working out and explain how you found the answers.

When you've finished, turn to page 80.

Sort the calculations

Challenge

Look carefully at these calculations.

Without working out the answers, sort the calculations into four groups. Each group must contain three related calculations.

56 + 38	121 − 35	12·1 − 8·6
1·267 + 0·582	10 − 3·83	12·67 + 5·82
38·3 + 61·7	1849 − 582	560 + 380
3·8 + 5·6	6·17 + 3·83	86 + 35

Once you've sorted the calculations, use mental strategies to work out the answers.

Explain the relationship between the three calculations in each group.

Think about ...

Think about how one of the calculations in each group can help you work out the answers to the other two related calculations.

For the 'What if?', each group of three calculations should include a mixture of addition and subtraction calculations, and calculations involving whole numbers and decimals.

What if?

Write a group of four related calculations (but not the answers).

Write a second group of four related calculations.

Rewrite your eight calculations, mixing them all up.

Give your eight calculations to a friend to sort.

When they have sorted the calculations, compare how you have both sorted the calculations and then work together to find the answer to each calculation.

When you've finished, turn to page 80.

Challenge

Each of these calculations involves adding or subtracting two 4-digit numbers.

What might the two 4-digit numbers be in each calculation?

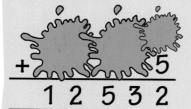

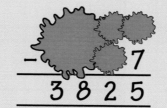

Explain how you worked out what the numbers are.

Now, for each calculation, write another two 4-digit numbers that give the same answers of 12 532 and 3825.

Think about …

You must use the ones digit shown in each of the above calculations, and the tens digit shown in the 'What if?' calculations.

Make sure that the two possible calculations for each answer involve using pairs of numbers that are different from each other.

What if?

Each of these calculations involves adding or subtracting two decimal numbers, each with 2 decimal places, for example 56·82.

For each answer, write two possible calculations.

Then explain how you worked out what the numbers are.

When you've finished, turn to page 80.

Challenge

Without working out the answers, use rounding to match each problem with its answer.

£4530 – £979   £945 + £3050   £1476 + £1899

£11 399 – £8254   £1265 + £1744   £16 750 – £12 999

£3375   £3551   £3995

£3009   £3751   £3145

Explain how you used rounding to determine the most sensible answer.

Think about ...

Which digits in the problem are you focusing on when estimating the answer?

Don't work out the actual answers to begin with. You need to estimate the answer first and explain how you estimated. It's your estimation that's important in this challenge, **not** the answer.

What if?

Which of these problems have an answer that is between £41.25 and £41.75?

£28.28 + £13.30   £69.05 – £27.90   £21.89 + £19.55

£62.75 – £21.39   £75.11 – £33.39   £25.36 + £16.44

Explain how you used rounding to estimate each answer.

Once you've made your estimations, work out the answers to check.

When you've finished, turn to page 80.

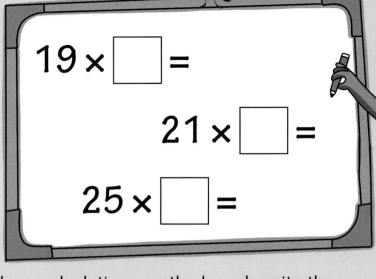

Challenge

$19 \times \boxed{} =$

$21 \times \boxed{} =$

$25 \times \boxed{} =$

For each of the three calculations on the board, write the same 2-digit number in the box.

Work out the answer to each calculation.

Do this five times, using a different 2-digit number each time.

How did you work out the answers?

Think about ...

Think about using factors, and remember the commutative law – that multiplication can be done in any order.

Think about using addition and subtraction as well as multiplication.

What if?

What if the three numbers in the calculations were 29, 31 and 35 instead of 19, 21 and 25?

What strategies would you use to work out the answers?

How can you adapt these strategies for multiplying other pairs of similar 2-digit numbers?

When you've finished, turn to page 80.

Challenge

I have just discovered that when you reverse the digits of a 2-digit number to make a new number and find the total of the two numbers, your answer is always divisible by 11.

Is Isabel's statement always true, sometimes true or never true?

Provide examples to justify your decision.

What happens if you use pairs of 3-digit, 4-digit or 5-digit numbers?

What do you notice?

Provide examples to justify your decision.

Think about ...

Use this divisibility test for 11:

- Add and subtract digits in an alternating pattern: starting from the left-hand end of the number and moving to the right, take the first digit, subtract the next digit, add the one after that, subtract the next and so on, alternately subtracting and adding the digits.

- If the resulting number is a multiple of 11, so is the original number. For this test, if you end up with a negative number, just ignore the minus sign and treat it as if it was a positive number. For example:

$$957 = 9 - 5 + 7 = 11 \checkmark$$
$$4635 = 4 - 6 + 3 - 5 = -4 \; ✗$$
$$3751 = 3 - 7 + 5 - 1 = 0 \checkmark$$

What if?

Isabel also says:

Investigate.

Provide examples to show what you discover.

I wonder what I might discover if I found the difference between the two 2-digit numbers?

When you've finished, turn to page 80.

39

Challenge

Replace each letter with a digit from 0 to 9. Identical letters must be replaced by the same digit.

The same digit cannot be used for more than one letter.

Some of the digits have already been given.

Write about how you worked out the value of each letter.

```
    A B 8 H
  ×       6
  2 2 B J H
```

```
      2 5 I
  ×       6 B
  C 5 5 H J
  C E C A
  C B A 5 A
```

```
        H 6
  8 ) A 6 8
```

Think about ...

Use your knowledge of the multiplication tables and also the inverse relationship between multiplication and division.

Using clues from one calculation will help you identify unknown digits in another calculation.

What if?

What if you replace each letter with a digit from 0 to 5? Once again, identical letters must be replaced by the same digit and the same digit cannot be used for more than one letter. But there is only one clue this time!

```
      3 V Y
  ×     Y W
  3 V Y V
  Y W V W
  Z W Y W
```

r does not represent a number, it strands for 'remainder'.

```
      Z W
  3 ) Y 3 W
```

```
      W 3 r 3
  Z ) U Y W
```

When you've finished, turn to page 80.

Write about how you worked out the value of each letter.

Challenge

$\frac{3}{5} \times 9 = 5\frac{2}{5}$

$\frac{3}{5} \times 9 = \frac{27}{5}$

$\frac{3}{5} \times 9 = 5\frac{4}{10}$

Who's correct?

Explain why.

Think about ...

You might like to include a diagram to help demonstrate your explanations.

Think about equivalent fractions, and equivalent mixed numbers and improper fractions.

What if?

Georgia also says: $2\frac{3}{4} \times 3 = 8\frac{1}{4}$

Write three other mixed number multiplied by a whole number calculations that have an answer of $8\frac{1}{4}$.

Explain what you did to ensure that each of your calculations had an answer of $8\frac{1}{4}$.

When you've finished, turn to page 80.

41

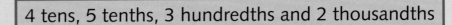
Challenge

Which of these expressions are equivalent to 4·532?

> 4 tens, 5 tenths, 3 hundredths and 2 thousandths

> $4 + \frac{5}{10} + \frac{3}{100} + \frac{2}{1000}$

> $4 + 0.005 + 0.03 + 0.2$

> 4 ones, 5 tenths, 3 hundredths and 2 thousandths

> $4 + 0.4 + 0.13 + 0.002$

> $4 + 0.5 + 0.03 + 0.002$

> 4 ones, 5 thousandths, 3 hundredths and 2 tenths

> $4 + \frac{5}{10} + \frac{2}{100} + \frac{12}{1000}$

> $4 + \frac{5}{1000} + \frac{3}{100} + \frac{2}{10}$

> 3 ones, 15 tenths, 3 hundredths and 2 thousandths

Write another expression that is equivalent to 4·532.

Think about …

Think carefully about place value.

Write different expressions using fractions, decimals and words.

What if?

Write three different ways of expressing each of these decimals.

0·76   **12·53**   **1·982**

When you've finished, turn to page 80.

Challenge

Although there are other factors that may affect what colour eyes you have, in general, you inherit your eye colour from your parents.

Parent 1		Parent 2		Likelihood of child's eye colour		
●	+	●	=	● 75%	● 19%	●
●	+	●	=	● 50%	● 38%	●
●	+	●	=	● 50%	● 0%	●

Look at the chart above. Work out the likelihood of a child's eye colour being blue.

Explain how you worked out the missing percentages.

Think about ...

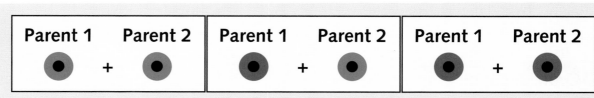

Show any calculations you've done – even if the calculations involved mental strategies.

What if?

Match each set of parents' eyes with a set of possible child's eye colour.

Parent 1	Parent 2		Parent 1	Parent 2		Parent 1	Parent 2
●	+ ●		●	+ ●		●	+ ●

Likelihood of child's eye colour			Likelihood of child's eye colour			Likelihood of child's eye colour		
● 0%	● 50%	● 50%	● 0%	● 75%	● 25%	● 0%	● 1%	● 99%

Explain how you matched the set of parents' eyes with the set of child's eyes.

When you've finished, turn to page 80.

Challenge

Write a fraction, a decimal and a percentage from those on the right in each of these statements.

$\frac{1}{3}$	$\frac{17}{25}$	$\frac{3}{10}$
$\frac{7}{10}$	$\frac{3}{100}$	$\frac{11}{20}$
$\frac{17}{100}$	$\frac{2}{5}$	$\frac{99}{100}$

0·35	0·8	0·09
0·67	0·14	0·73
0·28	0·92	0·41

58%	76%	39%
7%	12%	91%
20%	85%	43%

☐ < ☐ > ☐   ☐ < ☐ < ☐   ☐ > ☐ < ☐

How many different statements can you make?

Think about ...

Think about equivalences between fractions, decimals and percentages, for example $\frac{1}{4}$ = 0·25 = 25%, and how this might help you to compare and order them.

Make sure that each of your statements using the < and > signs contains a fraction, a decimal and a percentage. Try to use all 27 fractions, decimals and percentages in your different statements.

What if?

Order the nine fractions in the grid above, starting with the smallest.

Order the nine decimals in the grid above, starting with the smallest.

Order the nine percentages in the grid above, starting with the smallest.

Which set did you find easiest to order? Which set did you find the most difficult?

Explain why.

Now order the 27 fractions, decimals and percentages, starting with the smallest.

Explain how you ordered them.

Finally, look at your ordered set of 27 fractions, decimals and percentages. Arrange your ordered set into four groups:

- less than $\frac{1}{4}$
- between 25% and 0·5
- between $\frac{1}{2}$ and 75%
- greater than 0·75.

When you've finished, turn to page 80.

Challenge

Sort these measurements into three groups:

- length measurements
- mass measurements
- volume and capacity measurements.

0·4 kg	1750 m	$\frac{1}{3}$ litre	500 ml	18 cm
300 ml	1·2 kg	0·75 litre	20 cm	1200 g
$3\frac{3}{5}$ cm	0·5 litre	40 cm	0·25 kg	$\frac{3}{4}$ litre
$\frac{1}{4}$ kg	190 mm	$\frac{4}{5}$ kg	1·7 km	38 mm
0·8 litre	$\frac{2}{5}$ m	0·2 m	600 g	$1\frac{1}{4}$ litre

Then, using measurements from the same group, complete one each of these five statements for each group of measurements.

☐ > ☐   ☐ < ☐ < ☐   ☐ = ☐

☐ < ☐ > ☐   ☐ > ☐ < ☐

Choose one of your statements from each group and explain how you know your statement is correct.

Think about ...

Think about converting measurements to the same unit before comparing them.

Try to use each of the 25 measurements at least once in your statements.

What if?

Look at your three groups of measurements.

Order each group, starting with the smallest.

When you've finished, turn to page 80.

Challenge

What's the same and what's different about these measurements?

Sort the measurements.

inch

pint   kilometre

gram   **metre**   **litre**

pound   **mile**   **yard**   centimetre

foot   gallon   **kilogram**   ounce   millilitre

Explain how you sorted the measurements.

Think about ...

Try to sort the measurements above in several ways.

For the 'What if?', think about approximate equivalences between metric units and imperial units.

What if?

Carrots £2 per kilo

Carrots £2 per pound

1 pint £1.50

MILK

MILK

1 litre £1.50

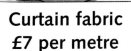

Curtain fabric £7 per yard

Curtain fabric £7 per metre

Look at each pair of items.

Which is the cheaper item in each pair?

Explain how you know.

When you've finished, turn to page 80.

Challenge

Look at this square.

The area that is shaded **red** is 9 cm².

The perimeter of the **red** square is 12 cm.

What is the length of one side of the **red** square?

The length of each side of the **blue** square is 2 cm more than the **red** square.

What is the length of one side of the **blue** square?

What is the area of the square that is shaded **blue** and **red**?

What is the perimeter of the **blue** square?

The length of each side of the **yellow** square is 2 cm more than the **blue** square.

What is the length of one side of the **yellow** square?

What is the area of the square that is shaded **yellow**, **blue** and **red**?

What is the perimeter of the **yellow** square?

The length of each side of the **green** square is 2 cm more than the **yellow** square.

What is the length of one side of the **green** square?

What is the area of the square that is shaded **green**, **yellow**, **blue** and **red**?

What is the perimeter of the **green** square?

Explain how you worked out the lengths of the sides of each square and the squares' areas and perimeters.

Think about ...

Make sure you show all of your working out.

What patterns do you notice regarding the lengths of the sides of the squares, their areas and their perimeters?

What if?

What is the area of the square that is shaded **blue** only?

What is the area of the square that is shaded **yellow** only?

What is the area of the square that is shaded **green** only?

Explain how you worked out the area of each of the three colours.

When you've finished, turn to page 80.

Challenge

In what years were Daphne and all her descendants born?

Explain how you worked out the year in which everyone was born.

Happy 97th birthday

Daphne Herne

31st December 2017

With love from your daughters,

Lorna (77), Joan (74) and Valda (69)

your grandchildren,

Ray (52), Wayne (50), Johnny (49), Jimmy (44), Marella (39) and Peter (37)

your great-grandchildren,

Alan (30), Daniel (28), Michael (25), Darren (23), Justin (22), Aaron (21), Allison (19), Christopher (18), Jessie (17), Blaze (12) and Craig (10)

and your great-great-grandchildren,

Taylah (4), Chloe (2), Lewis and Ryan (1) and Harry (5 months)

Think about ...

Use mental strategies to work out the year in which each person was born, and use known birth dates to work out unknown dates.

Show all your working out to help explain your reasoning.

What if?

Could your great-grandfather have been alive at the start of World War II?

Explain your reasoning.

Who in your family could have watched the first moon landing?

Explain your reasoning.

When you've finished, turn to page 80.

What's the new price?

Challenge

MINI DRONE

£40

WAS £40 NOW 25% OFF

> I worked out the new price like this: I know that 25% is the same as $\frac{1}{4}$ and $\frac{1}{4}$ of £40 is £10. So that means that the new price is £10.

> I did this to work out the new price: First I found $\frac{1}{4}$ of £40, because $\frac{1}{4}$ is the same as 25%. $\frac{1}{4}$ of £40 is £10. Then I subtracted £10 from £40, so the new price is £30.

Who worked out the correct new price?

Explain why.

Think about ...

> Think carefully about the word 'OFF' and what it means in this context.

> What operations do you need to carry out to find the new price?

What if?

Dave's Discounts

Roger Robot

£32

WAS £32 NOW 50% OFF

Robot World

Roger Robot

£24

WAS £24 NOW 25% OFF

EVERYTHING REMOTE

Roger Robot

ONLY £20

Which shop is selling Roger Robot for the lowest price?

Explain how you know.

What are the new prices?

> When you've finished, turn to page 80.

49

Challenge

Which of these are fake dice?

Explain your reasoning.

Facts

1–6 dice are the shape of a cube.
The opposite sides of a 1–6 dice always total 7.

A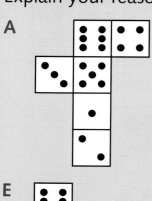

B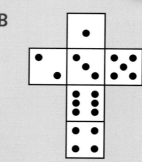

C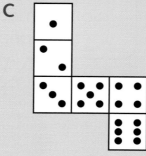

D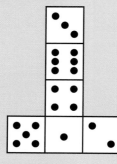

E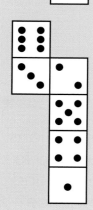

F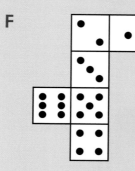

G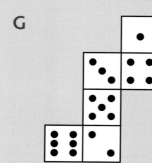

H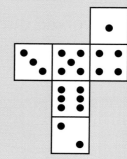

Think about ...

Think carefully about whether a net folds into a closed cube and also if the opposite sides of the cube will total 7.

Make sure that you explain why each of the above nets makes, or does not make, a 1–6 dice.

What if?

Can you draw a different net from those above that makes a closed cube?

Draw dots on your net so that it makes a 1–6 dice.

When you've finished, turn to page 80.

Challenge

Match each child with the angle they were calculating.

Explain your reasoning.

The unknown angle I calculated is 122°.

The size of the unknown angle I calculated is 118°.

I calculated that the unknown angle is 44°.

The unknown angle is 220°.

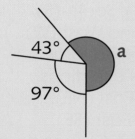
43°
a
97°

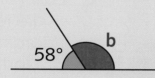
58°
b

117°
19°
c

d
152°

Think about ...

For each child, make sure that you explain how you know which angle they were calculating.

What if?

Name the type of angle that each child calculated.

When you've finished, turn to page 80.

Challenge

I'm facing North. If I make a clockwise turn to face East, I will have turned through 90 degrees.

What other statements can you make that involve making a turn through 90 degrees on the compass?

What statements can you make that involve making a turn through 180 degrees on the compass?

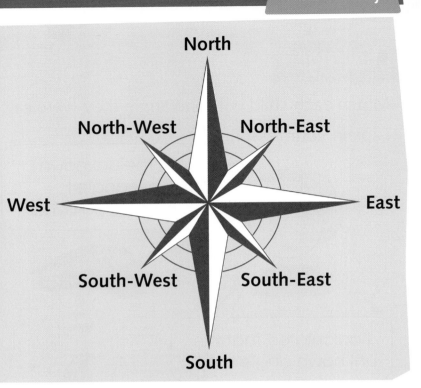

Think about ...

Think about making clockwise and anticlockwise turns.

Think about how many degrees you move each time you pass from one cardinal or ordinal direction on the compass to another.

North (N), South (S), East (E) and West (W) are referred to as 'cardinal directions'. North-East (NE), South-East (SE), South-West (SW) and North-West (NW) are referred to as 'ordinal directions".

What if?

What statements can you make that involve making a turn through 270 degrees on the compass?

What about statements that involve making a turn through 45 degrees?

When you've finished, turn to page 80.

Challenge

Mr Menel asked the children in his class to draw a symmetrical octagon on a coordinates grid.

These are the coordinates of the vertices of the octagons that Isabel, Joshua, Georgia and Xavier drew.

Isabel
(3, 9) (7, 9) (9, 5)
(9, 3) (7, 1) (3, 1)
(1, 3) (1, 7)

Joshua
(3, 9) (7, 9) (9, 7)
(9, 4) (8, 1) (3, 1)
(1, 4) (1, 7)

Georgia
(3, 9) (6, 9) (8, 7)
(8, 4) (6, 2) (3, 2)
(1, 4) (1, 7)

Xavier
(3, 9) (6, 9) (8, 7) (8, 2)
(3, 2) (1, 4) (1, 7)

You will need:
• first quadrant coordinates grid (vertical and horizontal axes marked to 10) or 2 cm squared paper
• ruler

Mr Menel says that only one of these is the set of coordinates for a symmetrical octagon.

Without drawing the four octagons, work out which child drew the only symmetrical octagon.

Explain why.

When you're sure you've identified the correct set of coordinates, use either a first quadrant coordinates grid with vertical and horizontal axes marked to 10, or draw and label a grid as shown below, and draw the octagon.

Think about ...

Look for patterns in each set of coordinates to help.

Explain why each of the children's sets of coordinates does or does not make a symmetrical octagon.

What if?

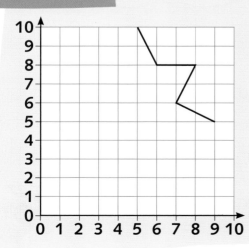

Mr Menel also asked the children in his class to draw an 8-pointed star that has both vertical and horizontal symmetry.

This is how Joshua started off his star.

Copy and complete Joshua's star and then write the set of coordinates for drawing the star.

When you've finished, turn to page 80.

53

A day on each of the planets

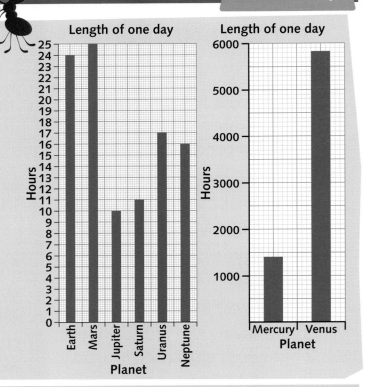

Challenge

One day is the time it takes a planet to spin around and make one full rotation. On Earth, this takes 24 hours.

These two graphs show the approximate length of a day on each of the planets in our solar system.

What statements can you make about the length of a day on each of the planets?

What statements can you make comparing the lengths of a day on various planets?

Think about ...

Given the scale of the graph on the right, you will only be able to make approximations, but make your approximations as accurate as possible.

You may need to use a calculator for the 'What if?'. Remember, 0·25 is equal to $\frac{1}{4}$, and $\frac{1}{4}$ of an hour is 15 minutes. Think about what 0·5 and 0·75 are equivalent to in hours.

Make statements comparing two or more planets.

What if?

To the nearest hour, how many Earth hours do you spend in school each day?

Approximately how many equivalent hours would this be on each of the other planets?

Think of other things you spend a lot of time doing each day, such as sleeping. Can you work out approximately how many equivalent hours this would be on each of the other planets?

When you've finished, turn to page 80.

54 Show all your working out.

The temperature of the planets

Challenge

This graph shows the average temperature on each of the planets in our solar system.

With the exception of Venus, the temperature decreases the further away a planet is from the Sun.

What statements can you make about the average temperature on each of the planets in our solar system?

What statements can you make comparing the temperatures on various planets in our solar system?

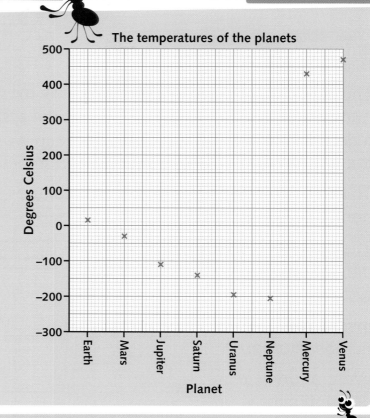

The temperatures of the planets

Think about ...

Given the scale of the graph, you will only be able to make approximations, but make your approximations as accurate as possible.

Make statements comparing two or more planets.

What if?

Earth, Mars and especially Mercury are known to have temperatures that vary quite widely.

	Earth	Mars	Mercury
Minimum temperature	−89°C	−153°C	−184°C
Maximum temperature	71°C	20°C	465°C

What statements can you make about the variations in the temperature on Earth, Mars and Mercury?

When you've finished, turn to page 80.

55

Challenge

The Hindu–Arabic numeral system is the most common writing system for expressing numbers as symbols in the world: it is the system that we use. It is also referred to as a decimal number system (from the Latin word decem, meaning 10) as it is based on the number 10 and uses 10 different numerals: the digits 0, 1, 2, 3, 4, 5, 6, 7, 8 and 9.

Throughout history, and in different cultures, there have been many different number systems.

The table shows six different number systems.

Invent your own number system.

Write or draw a description of your number system.

Decimal	0	1	2	3	4	5	6	7	8	9	10
Chinese	〇	一	二	三	四	五	六	七	八	九	十
Roman		I	II	III	IV	V	VI	VII	VIII	IX	X
Classical Greek		α′	β′	γ′	δ′	ε′	ϛ′	ζ′	η′	θ′	ι′
Ancient Egyptian		I	II	III	II II	III II	III III	IIII III	IIII IIII	III III III	∩
Babylonian		𒁹	𒁹𒁹	𒁹𒁹𒁹	𒐏	𒐐	𒐕	𒐖	𒐗	𒐘	𒌋

Use your number system to write today's date and your birthday.

Think about ...

Is your number system going to include a symbol for zero?

How are you going to express 2- 3- and 4-digit numbers? What about larger numbers?

Your symbols need to be easy to write, read and remember.

What if?

Choose ten different 2-digit numbers and write them using your number system.

What about choosing ten different 3-digit and 4-digit numbers?

Compare your number system with the number systems above.

What makes a 'good' number system?

When you've finished, turn to page 80.

Internet

Challenge

Investigate the amount of time pupils in your school spend each week on the internet.

On average, how long do pupils in your school spend on the internet?

Is there a relationship between their age and the amount of time spent on the internet?

Think about ...

Think about the best way to display your results.

Don't forget to include the amount of time spent on the internet at home and at school.

What if?

What do most people spend time doing on the internet?

Think about the different reasons for using the internet. These might include the following:

- 🖥 communication
- 🖥 social networking
- 🖥 finding out information
- 🖥 shopping
- 🖥 entertainment
- 🖥 business

Write about how much time people spend on the internet doing different things.

When you've finished, turn to page 80.

How are the reasons for using the internet different for children and adults?

Temperature differences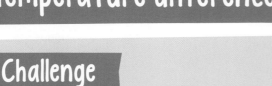

Using and applying mathematics in real-world contexts

Challenge

Look at each pair of thermometers. They show the average minimum and average maximum temperatures of five capital cities.

What statements can you make comparing the average minimum and average maximum temperatures in each city?

What statements can you make comparing the average minimum and average maximum temperatures between cities?

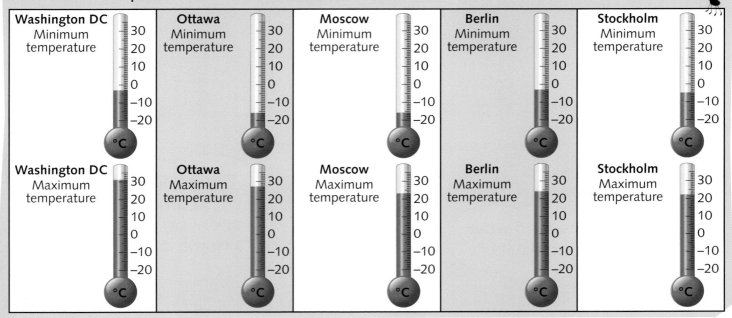

Think about ...

Think about temperature differences for each city and for pairs of cities.

Use the scales on the thermometers to help you calculate the temperature differences.

What if?

Which city has the greatest temperature difference?

How can you tell this without having to do any calculations?

When you've finished, turn to page 80.

Different nationalities

Challenge

If the parents of all the pupils in your class came to school for a meeting, how many different nationalities/countries would be represented?

What proportion of all the parents are from each country?

Which is the most common country?

Which is the least common?

Think about ...

Think about describing what proportion of all the parents and grandparents are from each country as a fraction or a percentage.

Think about the best way of displaying your results.

What if?

What if everyone in your class invited their grandparents to the school for Grandparents' Day? How many different nationalities/countries would be represented?

When you've finished, turn to page 80.

59

Domino magic square

Challenge

In a magic square the sum of each column, row and diagonal is the same: this is the magic number.

In the magic square below, eight dominoes are arranged so that each column and row (not diagonal) add up to the same total.

6	7	2	→ 15
1	5	9	→ 15
8	3	4	→ 15

↓ ↓ ↓ ↘
15 15 15 15 15

You will need:
• two sets of 6 × 6 dominoes

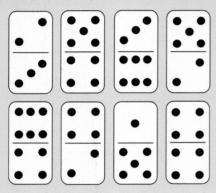

What other magic squares can you make using eight dominoes?

What's the magic number of each of your magic squares?

Think about ...

Try to create at least two different magic squares. Can you create one magic square using eight different dominoes?

Before you start, it might help to make a grid on which to place your dominoes, for example:

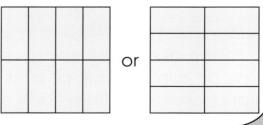

or

What if?

Show the position of any two of the dominoes in one of your magic squares to a friend.

Ask your friend to identify the six other dominoes and arrange them to make a magic square.

When you've finished, turn to page 80.

Domino calculations

Challenge

The two calculations below are made using dominoes.

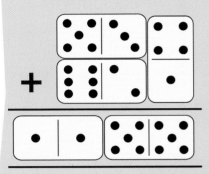

You will need:
- set of 6 × 6 dominoes

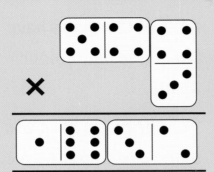

What other 3-digit add 3-digit calculations can you make using dominoes?

What other 3-digit multiplied by 1-digit calculations can you make using dominoes?

Think about ...

You must use dominoes as part of the answer to your calculation and you can't use the same domino twice in a calculation.

For the 'What if?', try to make fraction calculations involving non-unit fractions, for example $\frac{2}{3}$, $\frac{3}{4}$ or $\frac{2}{5}$.

Think creatively about how to arrange the dominoes, for example ▯∷ represents 6, whereas ∷▯ represents 60.

What if?

These dominoes show a fraction calculation:

When you've finished, turn to page 80.

What other fraction calculations can you make using dominoes?

Who am I?

Challenge

Describe yourself in relation to the other pupils in your class so that someone else can identify you.

You cannot describe yourself directly. That is, you cannot say your name, hair colour, eye colour, height and so on.

You have to describe how you are similar to, or different from, the other children in your class, but you can't name names!

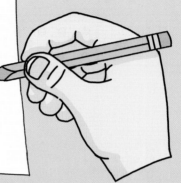

I have the same shoe size as 25% of the class.

45% of the class have the same hair colour as me.

$\frac{2}{5}$ of the class have the same eye colour as me.

I'm taller than $\frac{2}{3}$ of the class.

You will need:
- measuring equipment

Think about ...

Think about:
- shoe size
- eye colour
- hair colour
- height
- any other categories that are easy to measure.

Think about how you're going to collect the data necessary to be able to make comparisons and to describe yourself in relation to the other pupils in your class.

What if?

Who in your class would you say is most similar to you?

Why do you think this?

Describe why this pupil is so similar to you.

When you've finished, turn to page 80.

Burning calories

Challenge

A calorie is a unit of measurement – but it doesn't measure length, mass or volume. A calorie is a unit of energy.

Just as petrol fuels a car, calories are energy that fuels our bodies. You need about 2000 calories each day to keep your heart, lungs and brain functioning.

Every time you undertake a physical activity you use or 'burn' calories.

Activity	Approximate number of calories used per minute
sleeping	0·9
sitting	1·2
standing	1·6
washing and dressing	2·4
walking slowly	2·6
walking quickly	4·8
running slowly	4·4
running quickly	8·2

Use this table to calculate approximately how much energy you burn each day.

Show all your working out.

Think about ...

Don't forget there are 24 hours in a day. Have you accounted for all of them?

Remember, you are approximating how many calories you burn each day, so you will need to do some estimating and rounding.

The table shows the approximate number of calories used per minute for each activity. Think about how it might be useful to work out the number of calories used per hour or half hour for each activity. How are you going to calculate this?

What if?

You need about 2000 calories each day. Do you use more or fewer calories in a day?

Approximately how many more or fewer?

When you've finished, turn to page 80.

The triple jump

Challenge

hop **skip** **jump**

You will need:
- measuring equipment

The triple jump is an athletic event where athletes hop, skip and jump.

Work as a group.

Take turns to try the triple jump yourself.

Try three times and use your overall furthest distance.

Investigate what proportion of the total distance each of the three steps contributes.

Are the proportions the same for others in your group?

Think about ...

Think about how you are going to measure each of the distances: you will probably need to work in pairs or as a group.

Remember, percentages, decimals and fractions are different ways of expressing proportions.

What if?

The current male and female world record holders of the triple jump are Jonathan Edwards of Great Britain, with a jump of 18·29 m, and Inessa Kravets of Ukraine, with a jump of 15·50 m. Both records were set in 1995 at the World Championships in Gothenburg, Sweden.

What is the difference between your best triple jump and the world record holders?

When you've finished, turn to page 80.

You be the architect

Challenge

Plan your ideal house.

Draw a floor plan as accurately as possible.

You will need:
- squared paper
- ruler
- furniture catalogues

Think about ...

For the 'Challenge', think about:
- what rooms you want to include
- the number of floors
- the area of each room.

For the 'Challenge' and the 'What If?', think about:
- the size of each room
- which rooms go next to each other
- where the windows and doors are going.

For the 'What If?', there must be enough space around the furniture for people to move about.

What if?

As cities become increasingly overcrowded and the cost of land more expensive, architects are designing smaller and smaller homes.

Design a flat that has a total area of $30\,m^2$.

Include the measurements of each room and all the furniture.

Your flat must include:
- a bedroom for two people
- a bathroom
- a cooking area
- a living area.

When you've finished, turn to page 80.

65

News

Challenge

Investigate the amount of time that television channels devote to news programmes.

Approximately what fraction of time each day does each channel devote to news?

Does this vary from one channel to another?

Think about ...

Don't forget about current affairs programmes.

For the 'What if?', there may be some families who don't regularly receive news from any of the sources listed below, and some families may regularly use more than one source. Make sure you allow for this when collecting, organising and presenting your results.

What if?

The main ways that people receive news are through the following sources:

television **newspapers** **radio** **internet**

Find out which of the above is the most common way that families in your class regularly get their news.

Present your results in a table or graph.

What conclusions can you make from your results?

When you've finished, turn to page 80.

Using and applying mathematics in real-world contexts

Challenge

You will need:
• calculator

Calculate how much of your life you have spent taking a bath or having a shower. Show all your working out.

Think about ...

What unit of time are you going to use?

You can't possibly calculate an exact answer. You're going to have to base your answer on some assumptions. One assumption might be that the time taken for a bath or shower includes getting undressed, time spent in the bath or shower, drying off and getting dressed. What other assumptions do you need to make?

You will probably need a calculator to help you with the calculations needed to arrive at an answer. However, you must write down all of your calculations.

What if?

Mr Menel says:

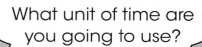

I've calculated that during my life I have spent the equivalent of 180 days in the bath or shower.

When you've finished, turn to page 80.

Could Mr Menel be right?

Shopping

Challenge

Look at a collection of supermarket receipts.

Choose suitable categories into which you can group the items on the receipt.

Do this for each receipt, using these same categories.

What trends do you notice?

You will need:
• supermarket receipts

Think about ...

Think about the different aisles and departments in a supermarket.

You need to have a 'reasonable' number of categories – not too many and not too few!

What if?

On average, which category of goods do people buy the most of?

Which category of goods do people buy the least of?

How are you going to measure this?

On average, which category of goods do people spend the most money on?

What about the least?

Is there a relationship between what people buy the most of and what they spend the most money on?

When you've finished, turn to page 80.

Shopping for the family

Challenge

You will need:
- supermarket receipts

Imagine you are responsible for next week's shopping for your family.

What will you buy?

How much will this cost you?

Think about ...

You need to make sure that you have enough food for everyone in your household (not just humans!) and that you provide a balanced diet.

Think about other things you might need apart from food.

What if?

What if you have £20 per person per week to spend?

Can you buy everything on your list with this? If not, what will you choose not to buy?

When you've finished, turn to page 80.

Plants in the playground

Challenge

You will need:
- online or printed gardening catalogues
- measuring equipment
- squared paper
- ruler
- coloured pencils

Your school has budgeted £300 to buy plants for your playground.

How would you spend this money?

Show an itemised account of how you plan to spend the £300.

Think about ...

Think about:
- the types of plants you would like to see
- how many plants you can afford
- whether the plants will last for just one season or longer
- where the plants are going to go
- other things you might need to buy, apart from plants.

What if?

Draw a design of how you would plant some or all of the plants you plan to buy with the £300. You can use either an existing garden in the playground or you can create a new garden.

Be as detailed as possible in your design, including any dimensions.

When you've finished, turn to page 80.

Mobile phones

Challenge

> I have worked out that on average, I make 3 phone calls a day from my mobile phone, each lasting about 4 minutes. I also send about 9 texts a day and I use less than 3GB of data each month.

You will need:
- mobile phone tariffs

Investigate the cheapest mobile phone tariff for Mr Menel.

Explain why you have chosen that tariff over other tariffs.

Think about ...

> As most tariffs are per month, think about what calculations you'll need to do in order to work out the most suitable plan for Mr Menel.

> Think about different mobile phone providers and tariff plans. Also think about whether it would be better for Mr Menel and his grandmother to go on a fixed contract or 'pay as you go'.

What if?

Mr Menel's grandmother has a mobile phone but she hardly ever uses it.

Mr Menel has worked out that his grandmother probably spends less than 1 hour a month on her mobile phone and she doesn't send texts or ever use the internet.

Investigate what plan Mr Menel's grandmother should be on.

When you've finished, turn to page 80.

Local currencies

Challenge

BRITISH POUND	GBP	1.000
US DOLLAR	USD	1.280
EURO	EUR	1.103
CANADIAN DOLLAR	CAD	1.621
HONG KONG DOLLAR	HKD	9.847
JAPANESE YEN	JPY	142.078
SOUTH AFRICAN RAND	ZAR	17.801
SINGAPORE DOLLAR	SGD	1.717
SWISS FRANCS	CHF	1.271
AUSTRALIAN DOLLAR	AUD	1.661

You will need:
- financial or travel section of a newspaper

Look at the exchange rates of different currencies against the British pound.

Choose ten different currencies.

How much local currency would you get if you changed £100 in each country?

In which country would you get the most of the local currency?

In which country would you get the least of the local currency?

Think about ...

Make sure that you show all of your working out.

For the 'What if?', you'll need to investigate the exchange rates against the pound over several weeks.

What if?

Look at the exchange rates of different currencies against the British pound.

Choose five different currencies.

Investigate how their rates have changed over time against the pound.

Do you now get more or less of each of these currencies for £100?

How much more or less?

When you've finished, turn to page 80.

Exchange rates

Challenge

This graph is called a **conversion graph**. It shows the exchange rate of British pounds to Euros.

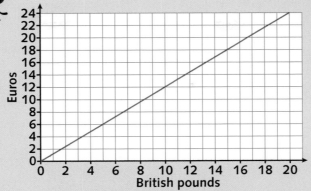

The graph is based on an exchange rate of £1 = €1.20.

Investigate the current exchange rate for the British pound against another currency, such as the US dollar.

You will need:
- financial or travel section of a newspaper
- graph paper
- ruler

Calculate the value of £10 in that currency.

Draw a conversion graph.

Think about ...

Think carefully about which currency to choose. Choosing a currency with an exchange rate to 1 decimal place (for example £1 = €1.20), and not 2 decimal places (for example £1 = US$1.35), will make it easier to draw and interpret your conversion graph.

Decide on the values you wish to show on the horizontal axis and the vertical axis. Start both axes from zero, and decide how far to number each axis and what scale to use.

What if?

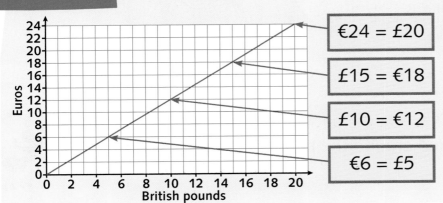

€24 = £20

£15 = €18

£10 = €12

€6 = £5

Use your graph to calculate the equivalent values of different amounts in both currencies.

When you've finished, turn to page 80.

Walking around the school

Challenge

You will need:
• measuring equipment

What is the perimeter of your school playground?

How long does it take you to walk around it?

If you were trying to keep fit by walking 2 kilometres every day, how many circuits of the school's perimeter would you have to do?

How long would this take you?

Think about ...

You will be working with different measures and with different units of measure. Think carefully about the relationship between the different measures and about converting between the different units.

Show all your working out.

What if?

When you've finished, turn to page 80.

How would this vary if you walked around the perimeter of the whole school?

All stand

Challenge

What is the maximum number of people that could fit into your school hall if they were all standing up?

Write about how you arrived at your answer.

You will need:
- measuring equipment

Think about ...

You will only be able to make estimates about the maximum number of people that can stand, lie or sit, but try to make your estimates as accurate as possible.

For the 'What if?', use your estimates for the school hall to help you with your estimates for the school playground.

What if?

What if everyone was lying down?

What if everyone was sitting on the ground, cross-legged?

Given what you have worked out for the maximum number of people that could fit into your school hall, can you estimate how many people could stand, lie or sit in the school playground?

If so, make an estimate and explain how you arrived at your estimate.

If not, explain why it's too difficult to arrive at a sensible estimate.

When you've finished, turn to page 80.

Painting

Challenge

Imagine you are painting the walls and ceiling of your classroom.

Work out the area of the walls and ceiling, and calculate how many litres of Pintu Paint you will need.

PINTU PAINT

5 litres is enough to cover 50 m².

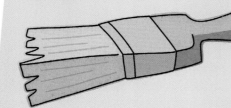

You will need:
- measuring equipment

Think about ...

Make sure you read the label on the tin of paint above. Use this information to help you calculate how many litres of Pintu Paint you will need.

Think carefully about the best way to find out the area of the ceiling.

What if?

Estimate first, then work out how many square metres of glass would be needed to replace all the windows in your classroom.

When you've finished, turn to page 80.

Making an octahedron

Challenge

This is the net of a square-based pyramid.

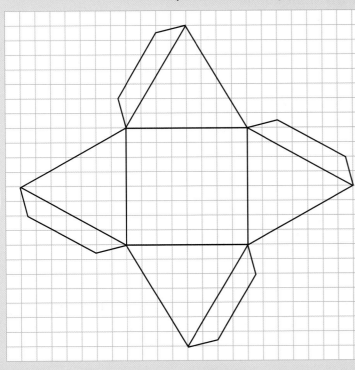

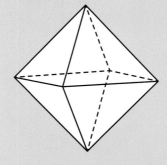

This is an octahedron.

You will need:
- squared paper
- ruler
- scissors
- glue
- sticky tape

Using the net of the square-based pyramid, can you make an octahedron?

Draw, cut out and construct your octahedron.

Think about ...

Use the dimensions of the net of the square-based pyramid above, that is a square with sides of 8 units and equilateral triangles with sides of 8 units.

Don't forget to include the tabs.

What if?

There are other nets that can be used to make an octahedron. Investigate what these nets look like.

Can you make another octahedron using one of these nets?

When you've finished, turn to page 80.

77

Find the places

Challenge

Choose a city near you.

Investigate how far, and in which direction, ten other places in the United Kingdom are from your city.

You will need:
- map of the UK (including scale)

Aberdeen

Glasgow • Edinburgh

Newcastle

Belfast •

Manchester

Liverpool

Birmingham

Cardiff

London

Bristol

Think about ...

Be as accurate as possible so that anyone following your instructions would arrive at the right place.

Think about the 8 points on a compass:

North

North-West / **North-East**

West — **East**

South-West / **South-East**

South

What if?

Look at the ten places you have chosen. Write them in order, starting with the place that is closest to your city.

When you've finished, turn to page 80.

Your favourite things

Challenge

My mum, football, chocolate, …

The seaside, crisps, reading, …

Swimming, playing the clarinet, bananas, …

You will need:
- squared paper
- ruler

Watching TV, karate, my little brother, …

What are your five favourite things? They can be anything!

What about the other pupils in your class?

What are the five most popular things in your class?

Think about …

There will be lots of different 'favourite things' and 'least favourite things' so think carefully about how you're going to categorise your data.

Present your data in a graph.

What if?

Xavier says:

I hate going to the dentist.

Isabel says:

I don't like broccoli.

Joshua says:

My least favourite thing is washing up dishes.

Georgia says:

I really dislike it when it rains.

When you've finished, turn to page 80.

What if you ask each pupil in your class to tell you their least favourite thing?

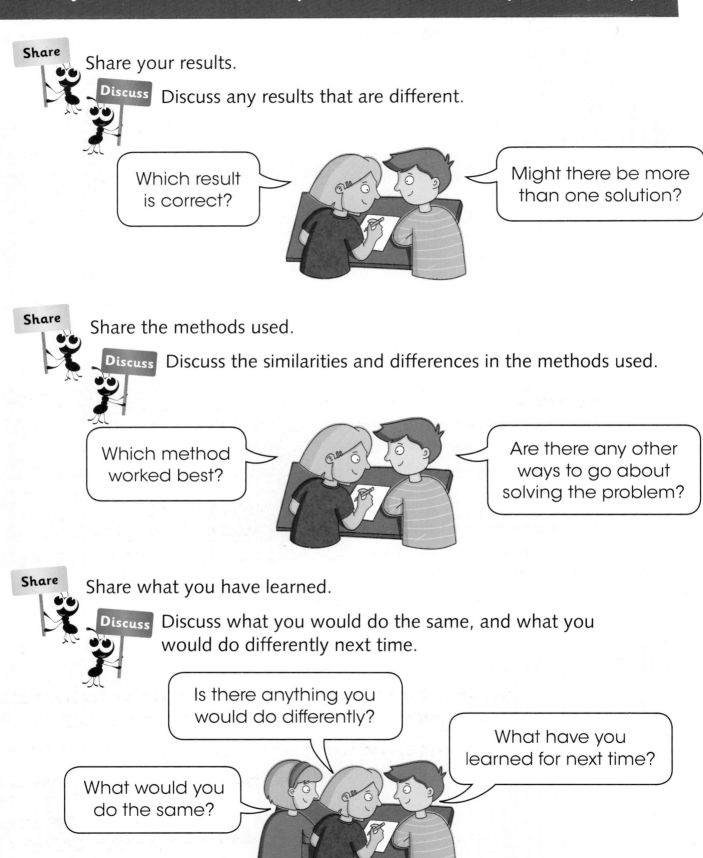

Share Share your results.

Discuss Discuss any results that are different.

Which result is correct?

Might there be more than one solution?

Share Share the methods used.

Discuss Discuss the similarities and differences in the methods used.

Which method worked best?

Are there any other ways to go about solving the problem?

Share Share what you have learned.

Discuss Discuss what you would do the same, and what you would do differently next time.

Is there anything you would do differently?

What have you learned for next time?

What would you do the same?